DATA SCIENCE JOB

HOW TO BECOME A DATA SCIENTIST

BY PRZEMEK CHOJECKI

ALL RIGHTS RESERVED @2020

© Copyrights by Przemek Chojecki, 2020

1st edition

Edited by Studio Grafpa

Table of contents

Introduction 7
Who is a Data Scientist? 9
    What is Data Science? 11
        Data Scientist 11
        Data Analyst 12
        Data Engineer 12
        Machine Learning Engineer 12
    Basic Concepts in Data Science 13
    What do Data Scientists do? 13
What skills you need to become a Data Scientist 15
    Communication and Visualization 16
    Technical skills you need 17
    From zero to hero 20
        Know your domain 20
        Base your models on data 20
        Build upon existing datasets 21
        Record your successes 21
        Success in data science builds your credibility 22
        Need to keep learning 22
Becoming a Data Scientist 23
    Getting an internship in data science 24
    Portfolio of GitHub projects 25
    Know what you know and what you don't know 26
    Polish your LinkedIn profile 26
    Shine! 27

Build a portfolio of open-source data science projects on GitHub ... 28
    Work harder, work smarter ... 28
    Start building a Github portfolio ... 29
    What makes a good data science project ... 29
    Experiment ... 30
    GitHub project ideas ... 31
        1. Clustering methods ... 31
        2. Scraping websites and extracting information ... 32
        3. Data Cleaning ... 32
        4. Data Visualizations ... 33
        5. Neural networks ... 33
    How to manage your Data Science projects ... 34
        1. Prioritize your tasks ... 34
        2. Figure out what your job is ... 35
        3. Keep a good documentation of your project ... 36
        4. Make sure that your team is always on the same page ... 36
        5. You need to think before you type ... 36
        6. dentify your most pressing research questions ... 37

5 steps to become a data scientist ... 38
    1. Find a company which needs data scientists ... 38
    2. Apply to data science internship programs and training programs ... 39
    3. Build your own open source project ... 39
    4. Join a local data science group ... 39
    5. Join a data science association ... 40

How to learn Data Science ... 41
    Data Science Courses ... 42
    Data Science Books ... 43
    Data Science Materials ... 44

Framework for learning and coding ... 46
    1. What you need to do ... 46
    2. Problem Statement ... 46

3. Write some code 47
   4. Prepare and evaluate your answers 47
   5. Review your answers 47
   6. Show your solution to the problem 48
   7. Identify Assumptions 48
   8. Identify Solutions 48
   9. Build a model 48
   Common mistakes Data Scientists make 49
      Trying to Learn Multiple Tools at Once 49
      Using Existing Code as a "black box" 50
      Lacking Consistency in Studying 50
      Not spending enough time on Exploring
      and Visualizing the Data 51
      Giving Tools and Libraries Precedence
      over the Business Problem 52
      Shying Away from Discussions and Competitions 53
      Not working on Communication Skills 54
   Conclusion 55

Data Science Books you should read 56
   Introductory level 56
   Intermediate Level 57
   Expert level 58

Should you do a PhD in Data Science? 59
   Is PhD for you? 59
   What is a PhD in Data Science? 60
   There is a life after PhD 61

You're ready 62

Data Science Crash Course 63
   Let's learn Data Science in 2020 63
   1. Introduction 63
      We use Python for Data Science 63
      Overview of Data Science Crash Course 63
      Getting and processing Data 64
      Standard techniques in Data Science 64

2. Anaconda and Jupyter Notebooks 66
   Jupyter Notebooks 67
3. Linear Algebra and Statistics 69
   Linear Algebra for Data Science 69
   Statistics 70
4. Processing Data 71
   Importing Data 72
   Storing Data 73
5. Getting Data 74
   Scraping web for data science 75
   Datasets available on the Internet 76
   Established datasets for Data Science 77
6. Classification and Supervised Learning 78
   What is Supervised Learning 78
   Classification and Regression Algorithms 78
   More Advanced Algorithms 81
7. Clustering and Unsupervised Learning 81
   What is Unsupervised Learning 81
   Clustering methods in Data Science 82
   Data Science is practical 84
8. Neural Networks 85
   What are neural networks 85
   Neural Networks in Keras 87
9. Dimensionality Reduction 89
   Principal Component Analysis 89
   Dimensionality Reduction in Data Science 90
10. Visualisation 91
    How to visualise in Python 91
    Share your open-source projects 94

Introduction

We're living in a digital world. Most of our global economy is digital and the sheer volume of data is stupendous. It's 2020 and we're living in the future.

Data Scientist is one of the hottest job on the market right now. Demand for data science is huge and will only grow, and it seems like it will grow much faster than the actual number of data scientists.

So if you want to make a career change and become a data scientist, now is the time.

This book will guide you through the process. From my experience of working with multiple companies as a project manager, a data science consultant or a CTO, I was able to see the process of hiring data scientists and building data science teams. I know what's important to land your first job as a data scientist, what skills you should acquire, what you should show during a job interview.

I will share this knowledge in this book, so that you too can join a wonderful world of data science - truly a market of the future.

If you'd like a quick summary of how to become a data scientist, here are 3 steps:
- Build your repository on GitHub and start an open-source project. You can take a dataset from Kaggle and build something around that. Usually classification problems tend to be easier. This will allow you to hone your skills and show a potential employer your engagement.
- Engage in Facebook and LinkedIn groups about Data Science and Machine Learning. Try to find meetups and conferences near you and attend them to meet more people. It's always good to have someone to guide you.
- Write more code! Data Science is a practical skill in the end. Share it on your social media — update your LinkedIn profile to have better chances to find a job.

From a technical standpoint you will need to know:
- fundamentals of statistics and linear algebra
- data visualisation methods (plotly)
- data processing and storing algorithms (pandas, NumPy)
- clustering and classification methods (KNN, k-means, DBSCAN, decision trees, XGBoost)
- how to apply dimensionality reduction (PCA)
- basics of neural networks (Keras)

That's it! I will discuss it all in detail in upcoming chapters. Also in the very last and long chapter I have included a Data Science Crash Course, which will give you the necessary theoretical background for entry-level data science jobs.

Let's start the adventure!

Who is a Data Scientist?

A Data Scientist is an expert with a deep knowledge of data, algorithms and data visualization. To be a data scientist, you need to possess the ability to work as part of a team, understand data structure, analyze data, design and create charts and graphs, and write concise code.

A Data Scientist used to be a scientific researcher who combines statistical methods and software development expertise in order to create, analyse, and interpret data. But now you can be a Data Scientist at a company, crunching data, trying to understand practical implications of real-world commercial data.

A Data Scientist is one of the most sought after jobs in the industry right now, and it comes with a lot of perks. You can earn anywhere between $120,000-$180,000 per year. For comparison, a Software Developer makes between $110,000-$135,000.

We're talking here specifically about the US market, but similar figures can be achieved in other developed economies. If you're not living in the US, Canada or Western Europe, you're still going to have a pretty large salary, in particular compared to what you can earn in other roles. This is again dictated by the fact that there are so few data scientists. On the other hand you can work remotely, as many US or European start-up-us are looking for remote help, which means no matter where you are, you can find a great job. The only thing you need is a good internet connection. Welcome to the global economy of 2020s!

Of course a Data Scientist's salary depends on their specific role, but they typically work in the field of analytics or machine learning, often working with large data sets.

They need to have:
- excellent analytical skills,
- experience in programming or databases,
- and strong writing skills.

A Data Scientist's job is to develop and analyze data to create actionable insights. These insights can be used in a variety of ways, including in various business decisions, and are often used to make recommendations or to help make a business case for a new product or service.

Data scientists work across a variety of different data science topics such as:
- business intelligence,
- web analytics,
- natural language processing,
- social media analysis,
- predictive analytics,
- machine learning,
- data mining.

The list really could go on and on. There so many applications of data science right now. So if you're thinking about becoming a data scientist, you can specialize in applying it to one industry of your choice (it's also fine if you don't have any preferences and just want to work with data). For example if you're passionate about people, you could go into ecommerce or marketing, to understand people's behaviour on a massive scale and help grow business at the same time. Maybe you're a fan of music, and would love to help Spotify or a similar start-up to grow their business. Everything is possible!

You can find a list of some of the key positions and how to apply to these positions on LinkedIn or AngelList or various other sites which allow you to browse job offers in your area (or remotely as I've mentioned).

What is Data Science?

Data science is a field of computer science which focuses on data. Data scientists combine methods such as statistical analysis and computer programs. Data scientists have a very specific set of skills that they use in their everyday work. The most basic job of a Data Scientist is to create data and then use software to analyze the data.

There are a few key types of Data Science jobs:

- **DATA SCIENTIST**

The most well known job title in Data Science. A Data Scientist is a very specialized job. Data Scientists are used in data analysis and modeling and are used to understand the complex world of data. There are a wide range of roles available for Data Scientists. Data Scientist positions can range from being a software engineer to having to write machine learning algorithms. They tend to be more on theoretical spectrum of Data Science.

- **DATA ANALYST**

A Data Analyst is a job title which is used for those that analyze large amounts of data to help companies make business decisions. Data analysts use tools like SQL to find patterns in large amounts of data to help them answer complex business questions. A data analyst job requires the ability to understand and communicate with data so they can understand how it impacts their business. They are between practical and theoretical aspects of Data Science.

- **DATA ENGINEER**

Data engineers typically work in business or business applications, which are responsible for developing systems, such as a database, that help people make sense of the data.

Some data engineers are engineers with a science or engineering background. This means that they may not have a computer science degree, but do have technical knowledge — a degree in mathematics or physics for example. They are the most practical when it comes to data.

- **MACHINE LEARNING ENGINEER**

Computer engineers and software developers who work with data processing software, have a very good understanding of the different approaches and approaches to data science. They are specifically oriented towards building machine learning problems which would be put to practical use.

That's at least a theory behind those roles. Practically you'll find on the job market each one of them mixed with one another, depending on the company you're applying to. That's why it's crucial to read the requirements for the job, or ask about it during an interview, because you might end up in a role which you haven't been applying to.

Basic Concepts in Data Science

Given that "data science" refers to numerous general topics (e.g. machine learning, recommendation, infrastructure, predictive analytics, etc.), let's give some examples. I'm going to take a narrow view, and look at most in terms of leveraging data science.

The most basic data science approach to problems is:
- Identifying the problem being solved: looking at a problem, asking what problems are related, how do we solve them and is there a technological solution?
- Explaining the problem: how can we measure the impact?
- Predicting the outcome: How many outcomes do we expect to see? How do we determine which outcome is likely to be the most positive?

What do Data Scientists do?

In an average day, a Data Scientist will look at raw data sets of any size, look at what it can do and make sense of it, use algorithms to analyse the data to build hypotheses, and test and validate the hypotheses with analysis tools such as statistical, machine learning, and data mining.

In an ideal world, a data scientist would be able to make sense of and analyse data at a level that can help the business solve complex problems and identify the best way to get to the bottom of the problem.

Summing up, a data scientist is an exciting role because it provides opportunities for growth and innovation. The job involves analysing data and making it accessible to users so they can make good decisions about the data. It involves understanding data formats, ensuring data security and privacy, and delivering quality solutions to customers.

What skills you need to become a Data Scientist

Data Science is exciting, rewarding, and fun. However, many people find the learning process overwhelming. Our jobs are expected to require a great deal of in-depth knowledge of technical data science areas.

Data Scientists are in high demand, so you'll need to have a well-rounded skill set if you're interested in this type of job. For a start, to become an intern or a junior data scientist you will only have to know:
- how to implement basic data structures in Python
- communicate and visualize your results well.

I will cover both those skills in detail in the last chapter in a form of Data Science Crash Course. The goal of this fast-paced course is to give you a theoretical background for any entry-level data scientist job. So altogether this book together with the course at the end will allow you to get your first job.

There's one but: you need to practice yourself. No coding, no data science. I'll get back to that in the course. Just remember:

data science is practical, and problem-solving is a mindset you can develop.

Now for some of non-technical skills you'll need as you progress along your career:
- Excellent analytical skills, including the ability to understand data
- Good project management skills, including the ability to plan and manage projects and communicate with others
- Good communication skills, including the ability to write clearly and concisely
- An understanding of the importance of data integrity and privacy issues, as well as the importance of knowing your users
- Excellent quantitative skills, including the ability to use data to gain insight and communicate the results clearly
- Ability to understand complex information and interpret results
- The ability to think critically and problem solve, both in terms of understanding the big picture and in terms of making specific decisions on how to solve a particular problem

Communication and Visualization

If you're thinking about honing your data science skills you should think about communication first. Communication is critical. Communication is a skill that is difficult to learn and become proficient in.

Communicating clearly, concisely and persuasively helps the user understand what you are trying to say. It also gives others a chance to explain their point of view and offer their opinion. Being thoughtful about the ways you communicate will help the audience have the best experience possible with your project.

Encouraging third party involvement, building relationships with your community is a valuable skill.

Then comes the trio:
- visualization of data
- visualization of statistical information
- data mining

In my opinion, the first two are more important for identifying potential problems, while the third is of utmost importance if you want to create great products.

To facilitate the developers, we also suggest to come up with an idea of how your visualization can be used to enhance your products. Often, it will lead to the development of a new tool, which can then be used on every step of the way.

Try to build helpful dashboards, explain, annotate and document well your projects! This way others can also use them to their benefit.

Technical skills you need

I need to say that first: **data science requires you to learn statistics and linear algebra**. And it's better if you do that before you start going into machine learning algorithms and do more advanced material. Without statistics you'd be often at a loss why certain methods don't work, you won't be able to interpret results properly and correct them in a good direction.

You'll be working twice as hard and long for something which is not necessary.

The other technical skill you need is coding in Python. You need to feel at ease with Python as you'll often need to prototype your solutions to a given problem. If you can do that quickly, you can then start iterating and arriving at a better solution. If that part slows you down and you need to think each time how to put your solution into working code, then you end up forgetting about the problem you're trying to solve by losing yourself in unimportant technical details.

Don't go this way — it's better to spend a couple of months polishing your statistics and Python skills than later having to return to the same material over and over when you'll need to solve a data science problem.

Finally, if you're accustomed to Python and statistics you should start polishing your problem solving skills. This is a never-ending story. After all Data Science is about problem solving. You need to be able to:
- pinpoint crucial data inputs,
- ask right questions about your dataset
- come up with new algorithms to process data
- extract information from data and use it later
- build complex solutions from small pieces

The best way to practice problem solving is... by problem solving. This is a practical skill. You can start coding and implementing algorithms to small problems, go on Kaggle and find a dataset/problem for yourself. It's all about finding new problems and trying to solve them.

Problem solving is a mindset which becomes a skill with enough practice.

You really have to program efficiently and quickly to do Data Science at levels higher than the most basic ones. And that's because Data Science is about testing, playing with data and experimenting.

You tend to have new datasets often and you need to be able to build a prototype quickly to at least research possible directions of where your work might be going. Without well learned Python you will lose hundreds of hours each time on simple implementation instead of thinking about algorithms.

From a technical standpoint you will also need some of the following — especially if you're applying for non-junior positions:
- Expertise in Excel, SAS, R or Python (note: Excel is a programming language)
- Experience in machine learning
- Experience with databases (e.g., relational databases or NoSQL databases)
- Experience in data visualization
- A strong technical background and/or background in computer science
- Ability to learn quickly

In addition, data scientists typically work in teams, which means you'll be asked to learn a lot of things quickly.

It's not all about data, but rather information, so you'll need to be an active learner if you want to develop the skills necessary to be successful in this industry.

From zero to hero

This subchapter discusses how does a data science career looks like and what it takes to become an expert.

- **KNOW YOUR DOMAIN**

Focusing on a single domain not only increases your intellectual ability but also gives you a rich reach in the wider field of data science.

You should try building your skills from a single domain to more vast subjects. Start with Python, go into Django, R, Java, learn something outside your comfort zone. This way, you'll also get better at techniques you use on a daily basis. It's all about practice.

- **BASE YOUR MODELS ON DATA**

Instead of just playing with models to fit a numbers model that the machine learning community uses, you should also build upon what has already been done by the community.

If you are looking for formulas to model for variables of your domain, the task will become a lot easier if you know all the formulas available in the community.

- **BUILD UPON EXISTING DATASETS**

In this field, it's a matter of either building on existing data sources or building your own datasets.

Building on data that has existed for a long time, as opposed to building your own datasets, will give you the proper feedback for your models which will help in the formulation of your models.

Focusing on the core strength of data science is tackling the issues around the core problem. Data science offers you unique opportunities to apply what you are learning to real-world use case problems. And by taking advantage of existing data, you will be able to validate your new knowledge.

- **RECORD YOUR SUCCESSES**

Recording your successes is also a must, as you fail hard and learn.

Record the things you've done and any learning that you've done.

This will help you when needed to replicate your work.

In addition, data science is mainly about the iterative development of the methods that we've learned. Code, reiterate, code again.

You need to validate all of your methods at least 3 times.

By recording your successes and failures, we will be able to look back to your work and put it in context in the community of data scientists.

- **SUCCESS IN DATA SCIENCE BUILDS YOUR CREDIBILITY**

To be credible in the industry, you need to do something at least as well as everyone else.

If you find the data science market too competitive, then raise your standards by focusing on excellence, making your way up the high paying data science market.

That way you can become successful from this, and that will also motivate other data scientists to innovate.

- **NEED TO KEEP LEARNING**

Data science is a very fast-paced industry with lots of different challenges thrown at you in every step. This makes it easy to stagnate or fall behind in your life. This is because you have to keep learning to keep up with the competition.

Maintaining a high learning pace is also imperative to keep in on top of your competition. This will help you attract better jobs and enhance your brand.

Becoming a Data Scientist

This is a great time to learn about data science. You can start your own business as a data scientist, you can start building a career, and you will make a good living. I have already talked about how to get into data science, and it's time to learn how to do it well. This chapter is about practical things you need to do to become a data scientist.

It will take some time, but you will get there.

When I first started out, I started out by reading all the books on the subject, but I soon realized that many of them were out of date.

I now prefer to get my hands dirty with real data.

If you want to start a career as a data scientist, it's time to start collecting data.

There is more information and tools that are now available to help you gather data than ever — starting with Kaggle for public databases, GitHub for open-source code, Reddit for many links and articles you can scrape and re-use.

Getting an internship in data science

Of course you don't have to start a business to become a data scientist. Another natural way is to find an internship. There are plenty of ways to get internships, but the first way is to simply contact a recruitment company - just search on LinkedIn for recruiters in your area and write to them. They will be often willing to help you.

The second way is to seek out companies that want to hire interns and reach out to them to see if you could get an internship. Cold-calling really works if you can show your work (GitHub again!). Check websites of companies to see whether they are looking for data scientists or ask someone already working in a data science team.

The more proactive you are, the better. Start writing about your data science projects on LinkedIn and publish posts on Medium to gain more visibility. The perfect situation is when you're approached by recruiters and are the one to choose. The only way to do it is to put in the work into your open-source data science projects and popularize what you're doing.

If that's not the way for you — because you simply don't feel comfortable with being so public, which is perfectly fine — just start looking for internships on many job websites and groups. It's an employee market and if you can show your engagement and passion, you should be able to get an initial call or meeting from a potential employer.

Portfolio of GitHub projects

First of all you should build a portfolio of open-source projects on GitHub. I recommend to create three projects:
- A classification project where you use a public database (you can download one from Kaggle — more about that in the next paragraph) of images/texts to sort them and hone your skills with supervised/unsupervised learning (from PCA to neural networks, through DBScan, KNN, etc.).
- An NLP project which would analyse sentiments from Tweets from a particular subject and classify them accordingly into positive/neutral/negative. That's a classic one — choose a topic interesting to you, so you have a good story to tell about it.
- A scraping project, where you scrape different sources to extract information — that can be scraping sports news if you're a sports fans, financial fluctuation if you're into finance or data science novelties. The end goal might be creating an automated website with scraped and extracted content. Easy to show during a job interview.

If you're more advanced then you definitely should play around with latest machine learning algorithms. For example you can try:
- GANs (Generative Adversial Networks) and generated some faces or cats.
- Reinforcement Learning with easier games.
- GPT-2 and text generation.

You have endless opportunities here. You definitely should show that you're well-versed in neural networks and their fundamentals like Keras, PyTorch, TensorFlow.

A **great resource is Papers with code** (Google it!), which links machine learning papers with Github code. That's a very efficient way of finding most recent findings in machine learning.

Know what you know and what you don't know

Even if that sounds obvious you should be able to answer questions about your knowledge, particular technicalities or problems.

Actually the most important question you'll be asked is what was the hardest in a given project. If you were struggling with a particular technical issue, why and what it was. How did you overcome it? Those are the basic kind of questions that you should be prepared to answer.

Be ready to talk about algorithms in more detail, different methods you have used in the past. Be open and share also what was problematic for you. This can only help during a job interview.

Polish your LinkedIn profile

The last thing which is often overlooked by data scientists is putting up a coherent LinkedIn profile with explanation of what you did in the past and where you're at currently. If you have any gaps in your career be sure you'll be asked about them — there's nothing wrong with 6-months stay on Bali surfing, you should just be honest about what motivated you back then and what you wanted to achieve. I imagine that a good argument would be you wanted to have a break

from your screen, wanted a restart, wanting to work as a digital nomad — there are plenty of reasons to explain why this option was the best also for your career.
You should also ask for recommendations from your past employers. They can write it directly on LinkedIn — just a couple of sentences is enough. If your last departure wasn't really planned and you feel bitter about it, try to explain why and what you expect from your next employer.

You should be able to say precisely why you want to change your job. Is it because you're looking for new challenges? If yes, why you can't have them at your current job? Speaking clearly about your previous work experiences is a huge asset.

Shine!

Summing up there are 3 things you should do to really master a job interview for a data scientist position:
- build a portfolio of GitHub projects;
- know what was the hardest part of each of your project and how you overcame it;
- polish your LinkedIn profile.

I'm a huge fan of building your own portfolio of open-source GitHub projects as a way to grow as a Data Scientist. In the next chapter I'm going to discuss a list of projects you can try for yourself if you're into practical data science. Those will definitely get you hired!

Build a portfolio of open-source data science projects on GitHub

'Science' in Data Science can be misleading. After all Data Science is a very practical endeavour. You need to code a lot in order to really get it. Let's discuss how to get practical.

Work harder, work smarter

So you want to be a better Data Scientist? Code. Gather databases and visualize, analyse, understand it. And then learn how to apply that knowledge to new data sets.

A big part of being a Data Scientist is to make sure you're doing all your analysis on as much data as possible, because data is your biggest asset. No matter how clever algorithms you're using, if you don't have enough data, you won't progress.

You can always go back and try to get your best set of results again, but as a Data Scientist you really want to get as much data as you can before you start looking at it in a more systematic way.

Start building a Github portfolio

The sooner you start building a Github portfolio of open source projects the better. This way you'll not only learn a lot, you'll also get a better chance at finding your next dream job. It hugely increases your visibility as a Data Scientist and allows you to document your Data Science journey.

If you're interested in working in the Data Science and you're not sure which company to choose, try and get a quote from a company that is in a very specific niche or category. It'll be easier to convince the recruiters that you can fill a specific job requirement rather than being seen as a general Data Scientist.

Practically speaking the best way to do that, would be to research the companies in a particular niche — for example through Crunchbase — and then build a portfolio of Data Science projects related to that niche. If that's niche is your hobby as well, it's a win-win situation.

What makes a good data science project

If it is something you think you can do, or something that you think is a good fit with the data science mindset, and you have the chance to learn the skills and experience needed to do it, take it!

Data science projects are all about learning, and the better the data science skills you have, the more likely you are to build a successful project.

The following is a list of the most important aspects to consider when creating a data science project.
- The project needs to have some sort of measurable result.
- The project should be simple.
- The project should be well understood by the client.
- The project should be in the client's control.
- The project should be in their best interest.
- The project should provide clear, concise information and results.

You should start a project with the data. It is often the first step. Start with what you know, what you don't know, and the reasons why.

You should have a clear project agenda and timeline.

The project must start with a clear idea of what you are trying to do. If you don't, you will not have a clear direction. That's often fine at the start, but with time you should plan your work to get the best results.

Experiment

Data Science is all about experimenting. A good tool can give you a better understanding of what your data looks like, but it's only useful if you can extract insights from it. When looking at a dataset, you need to be able to do a lot more than see if it has a pattern.

To get insights from a dataset, you need to know what to look for. What are the patterns that you can identify and how can you turn that into actionable insights?

If you've got your head around these questions, then you can start to get insights that can lead to business outcomes. You can start to use these insights to start to improve the quality of your work and work on your skills.

Once you've got a solid understanding of what you're looking for, there are many ways you can go about getting insight from your data.

Finally, one of the best ways to look at your data is by creating your own report. Have a spreadsheet where you note everything you did in a project. This way you'll keep everything well organized, you'll learn a lot and be ready to tackle more complex projects.

GitHub project ideas

Let's finish this chapter with five GitHub projects you may try to do for yourself before you approach job interviews. The following 5 ideas will let you build a varied and fun portfolio of projects:

1. CLUSTERING METHODS
One of the first things you learn after you learn the basics of Python is how to classify data and put them into groups relative to their different features. You can classify text according to sentiments or images according to similarity. You can go for Kaggle's different datasets to classify sports' top performers and try to determine what's making an athlete a star.

Options here are endless and you should definitely go into playing data as much as you can.

As for the methods, you'll be using definitely DBSCAN and/or KNN to cluster your data. Also, some dimensionality reductions, like PCA or t-SNE, might be helpful. I will discuss those methods in the last chapter.

2. SCRAPING WEBSITES AND EXTRACTING INFORMATION
What's more fun than choosing a website which you visit often and try to scrape it? You can go for financial news, social media trends or best cooking recipes. You can even try to scrape Google and other search engines — it's not hard in Python.

Scraping a single website is fun because you get to understand how websites are built and their flow of information, from a totally different perspective.

Scraping search engines is great for learning how different pieces of information are spread throughout the internet.

And if you connect the two — you'll get an amazing project: newsfeed about a particular subject!

Technologically, you'll be using Requests library to scrape and BeautifulSoup to clean the html code. Alternatively, you can also use Selenium to scrape.

3. DATA CLEANING
Here's a standard though a little tedious task: cleaning data. It is standard because this happens to be a part of almost any real-world problem you will encounter as a data scientist. It is tedious because you might end up spending long

hours thinking about how to make sense of the enormous data you have.

Data Cleaning might involve taking ticks of stock prices and trying to make sense of that — by clustering, by good visualization. Or if you already have a data set, like your old documents and photos, it might involve trying to tidy them up, by deleting those which are not needed anymore.

Data Cleaning might be as easy as deleting unwanted files to as hard as trying to use machine learning to spot anomalies in your data set to delete those.

One of the standard libraries here is pandas and NumPy.

4. DATA VISUALIZATIONS
Whether you have a nice data set or not, the best way to understand data is to try to visualize it in many different ways. It's not only helpful for you, but also for your potential clients or employers.

By visualizing it might be much easier to spot what's going on.

Here Plotly comes in handy, which is a standard Python library for plotting functions and graphs. If you want to build an interactive dashboard then you should try using Dash in Python.

5. NEURAL NETWORKS
Time to get into machine learning! Neural networks are not as hard as you think. Their easiest instances are very quick to

implement, especially if you use a framework of Keras, which does a lot of work for you.

The entry point for neural networks is classifying hand-written digits. If you Google "MNIST Python" you'll find plenty of tutorials on how to use simple neural networks to do it.

From then you can go into more complex datasets (e.g. images of animals) and problems (multi-class classification, anomaly detection).

How to manage your Data Science projects

One of the hardest things about this job is managing your data science project. Organize your desk and work environment. Schedule your tasks and prioritize. You probably have to think through all of your options for how to approach every aspect of your data science project.

Here are a few things you should keep in mind as you move through these steps:

1. PRIORITIZE YOUR TASKS
When I was first in the industry, I would spend a lot of time looking at what was next on the data science agenda.

After a few months, I realized that this was not a very productive approach. What was really helpful was to focus on a few key data science tasks at a time. When you do this, you're going to be able to complete your work faster. In fact, you will have much more clarity in what you need to do next.

You can run the following process:
- Pick the data you want to analyze
- Identify the research question(s) you want to answer
- Define the type of analysis you want to perform
- Pick a research plan
- Create a data pipeline
- Write your code
- Repeat

Once implemented, you can run it over and over, to get better each time.

2. FIGURE OUT WHAT YOUR JOB IS

The first thing to do when building a data science team is to figure out what your job is.

If you're working in a startup, it's probably pretty clear — you need to build something useful and interesting that people can use.

But in your company you might need to do some pretty fancy analytics stuff, like:

- Using data science to build a better product (or a better solution) for your customers
- Improving your marketing and selling strategy
- Driving ROI for your business

The trick is to figure out which one of these things you actually need to focus on and to do right now. So we're back to prioritizing.

3. KEEP A GOOD DOCUMENTATION OF YOUR PROJECT

Even when you know what your business does, you don't always know what your next steps will be and how long it will take. You may need to start again, or you may have new tasks you don't even know about.

I keep an Excel spreadsheet (and Trello and Evernote) for these things:

- What I need to do today
- How long until I can start
- What else needs to be done today
- What are the next steps that need to be taken

4. MAKE SURE THAT YOUR TEAM IS ALWAYS ON THE SAME PAGE

Every time you make a change, you need to make sure that everyone is on the same page. A good team is built on collaboration. If everyone is working in sync, your company is going to be better and faster.

Keep it simple.

5. YOU NEED TO THINK BEFORE YOU TYPE

If you make a mistake, you can try to make it up later. Or, you can make it up now and move forward.

We have all made mistakes in our lives. It's part of the learning process.

You need to be aware of what the other person is typing in order to avoid a mistake.

6. IDENTIFY YOUR MOST PRESSING RESEARCH QUESTIONS

As you move through this process, you will see the different challenges of your research and you'll want to identify the questions that need to be answered most.

Data Science is practical. It's all about how you handle data and what insights you can gain from it.

So when you need to make decisions, it's important to take the time to consider the data that you've already collected. It's a good idea to keep a few key documents around to help you make decisions.

5 steps to become a data scientist

Let's now focus on actually finding a Data Science job. There a couple of steps you might take to locate opportunities right for you in any country.

1. Find a company which needs data scientists

Companies will often hire data scientists for specific projects. For example, a financial institution might have a company-wide project where they want to build an algorithm for analyzing data.

If you're interested in learning about the types of companies hiring data scientists, go on AngelList and start going through data science jobs in startups.

Remember you can both look locally but also globally, because many companies are willing to hire a data scientist to work remotely for them. AngelList supports this type of remote jobs, so have a look there.

Another option is looking at Crunchbase to see Artificial Intelligence or Data Science companies in your location.

Even before your first job you might want to use Fiverr to get your first gig. Fiverr is a perfect way to start earning as a Data Scientist, no matter who you are and where you live. All you need is a connection to the internet and then a good, complete profile so that people can see you. I'm still using Fiverr myself from time to time to find interesting consulting roles from around the world.

2. Apply to data science internship programs and training programs

Internship programs and training programs are a great place to get a glimpse into the industry and learn a new skill.

Reading a data science book can help you gain a deeper understanding of the industry. In later chapters I review most useful Data Science books.

3. Build your own open source project

While data science is becoming more and more important in the industry, there are still tons of data scientists that are not part of the larger open source community.

There are tons of free tools and resources on GitHub, and plenty of data scientists are eager to share their knowledge and contribute back to the community.

4. Join a local data science group

Some cities have local data science groups that are dedicated to helping members. While it's important to try to stay up to date with the latest tools and techniques, this community is where you can learn about data science as a whole and how it fits into business.

Each larger city now boasts an active group of data scientists.

5. Join a data science association

Data Science can be very intimidating and it can be difficult to start learning about different techniques, but that doesn't mean you can't do it.

Here are just a few things to consider before jumping in.

First and foremost, it's important to get involved in data science discussions and community-driven discussions to understand what data science is, where it is heading, and how best to leverage it. Reddit is great for that, as well as various Facebook groups.

A good place to begin is by attending data science meetups and data science conferences. These meetups provide a space where anyone can learn about data science, be it the basics or advanced techniques, as well as provide the tools and training to get started. If you don't have a specific project in mind you can look at the events list to see who is hosting the meetup and if there is a specific time when it is happening. After you finish looking through the meetups list I would recommend trying to connect with the organizers. Attend any conferences that they will be attending and ask them if you can attend their next data science meetup. You can also follow the organizers on social media to see what's going on.

How to learn Data Science

We can divide sources of learning data science into 3 groups:
- data science courses
- data science online materials
- data science books

Firstly I should mention that I'm assuming that you've already know the fundamentals of Python — at the level of being able to import different data formats and clean them in a Jupyter Notebook. If not, read the Data Science Crash Course at the end of this book, it's for you!

Secondly you should know basics of statistics and linear algebra.

If you don't know any of these, the I recommend you start with:
- CodeCademy to learn Python for free.
- For online and free source of statistics check out Think Stats for a longer treat. You can also find many posts online which cover quickly how to manipulate with matrices, probabilities or other fundamental mathematical concepts.

However you probably would like to spend more time on those fundamentals if you're just starting and the best for that would be taking courses (as a free reader/listener) at your local university or follow a curriculum of one of established universities.

Data Science Courses

There are many courses and organizations that have been established to help people to get started. Google things like 'Data Science Bootcamp' or 'Data Science for Everyone' and you'll find courses specifically meant to help people to get started with data science in Python. The offline courses can be quite expensive. But if you are ready to jump in and start learning, I would recommend you spend a couple of hours a week to take the course online and you will definitely be more comfortable. If you decide to go a the data science bootcamp, I'd recommend that you first work on a real project that you can show. This will give you a good idea of what data scientist does.

If you want to complete courses in machine learning and data science, the first one you've probably heard about is the course by Andrew Ng:

Machine Learning by Andrew Ng on Coursera (https://www.coursera.org/learn/machine-learning) — it covers fundamental concepts of machine learning, is a bit more theoretical (less coding involved) and is a fantastic source of what to know. However if you're really just starting there are other sources you should check.

So here's the list of more basic courses that you should take at the beginning of your data science adventure:

Harvard Data Science course (http://cs109.github.io/2015/pages/videos.html) — takes you through basics: pandas, scraping, regression to machine learning.

Introduction to Data Science in Python by University of Michigan (https://www.coursera.org/learn/python-data-analysis?specialization=data-science-python). Highly rated, basic introduction. It's on Coursera so it's free if you don't need a certificate for it.

Deep Learning (https://www.coursera.org/specializations/deep-learning) — though it's far from introductory, this course is another one from Andrew Ng you should at least hear about.

IBM Data Science (https://www.coursera.org/professional-certificates/ibm-data-science)— one of the most popular data science courses on Coursera right now. It is offered by IBM and offers a complete entry into Data Science from scratch. Especially great if you want to have a certificate for completing a course.

Data Science Books

I'm always trying to read a lot and books I list below are some of my favourites. For a more complete list, see one of the last chapter of this book.

Data Science from Scratch: First Principles with Python — this is probably one of the first book I've read about Data Science. It really starts with basics and do not assume any knowledge of Python. Great for absolute beginners!

Python Data Science Handbook: Essential Tools for Working with Data — more advanced book which will lead you through different data science techniques.

Introduction to Machine Learning with Python: A Guide for Data Scientists — another introductory book, this time focused more on Machine Learning.

Data Science Materials

I'm a huge fan of GitHub. Well, like almost everyone. A really great source to learn data science is actually through browsing GitHub's countless open-source repositories.

The way to browse it is to search for a given concept and add Github on Google. For example "data visualisation github" or "DBSCAN github". You can also browse Papers with Code (www.paperswithcode.com) to see the most recent research done with a code you can see and use.

This way you'll be able to quickly find working code that you can play around with on your computer.

Another thing is data science conferences. You should definitely look for some in your area and attend at least one to see what data scientists talk about and what is trending right now. This is a great way to keep up with the latest tools and techniques in this field and learn from the best data scientists in the industry. If you like to stay updated with what the data scientists in the field are writing, keep an eye on the Data Science subreddit and the blogs of the best data scientists. And make sure to keep up with all the latest tools and techniques the data science community is developing in the world of big data. This will ensure you have access to the latest tools and techniques that will allow you to leverage your experience and skills.

Framework for learning and coding

It's not always easy to become a data scientist, even for the best of engineers. However, it's not too difficult for any student willing to work hard and there are a few things you should know beforehand. This chapter takes a look at them, providing you a clear road map to succeed in data science and giving a framework to learn anything. This will also help you approach coding problems especially during a job interview.

1. What you need to do

- Set a target number of hours
- Plan what you need to know/experience in order to write code
- Pick an interesting dataset and perform a search
- Acknowledge that the biggest challenges are in the start
- Learn how to ask the right questions

2. Problem Statement

For every question, write the problem statement in plain English.

This is also a good time to take a look at some technical examples of the problems that real companies encounter with the data science process.

3. Write some code

Make your code and run it. The interviewers will take a look at the code and answer the questions afterwards.

- Estimate how much time you will need for the data preparation
- What should be an average number of people on your team for solving the problem?
- What can go wrong?
- Write down the tools that you will use to tackle the problem

4. Prepare and evaluate your answers

Make sure to have valid solutions that are both concise and creative.

Write down your responses to the questions. Make notes. Show engagement.

5. Review your answers

Refine your answers, ask yourself why you didn't make use of some existing frameworks or tools.

This will help you identify areas that you need to learn more about, and convince yourself to write up some material for future interviews.

6. Show your solution to the problem

As a bonus, prepare and show your solution to a third party who will help you get in touch with the organization that interviews candidates.

Prepare a notebook with all of your solutions so you can discuss your solutions with interviewers.

7. Identify Assumptions

Define the assumptions you think your interviewer has in the company and the tasks that you will be dealing with. Do not compromise on these assumptions as it's a good way to separate yourself from the crowd.

8. Identify Solutions

Identify the solution that meets your interviewer's expectations. Ask questions, be open and honest, try to understand your interviewer's perspective.

9. Build a model

Every question has a possible answer that fits the data you have. Invest your time in defining a robust solution, applying it, and prove it during an interview.

The experience in programming is crucial for data science: You will have to execute the code and find bugs. The faster the better.

If you are nervous, you can also ask questions in the interview about how to structure your code to fit with the structure of the company, and how to track what work you need to do for the upcoming weeks if there are multiple parts of a job interview.

Common mistakes Data Scientists make

Let's review 7 common mistakes which people make along the way. Thinking about those will help boost your career and learn much quicker.

- **TRYING TO LEARN MULTIPLE TOOLS AT ONCE**

The result is that we become paralyzed with an inescapable set of tool-specific knowledge that we do not have any ability to leverage on a broad level. This is why so many programmers spend time learning and practicing new skills on a particular tool at a particular moment, when they have little or no ability to adapt and adapt quickly to new situations in the real world.

A common analogy is a person who tries to learn a new sport by learning a lot of different skills in a short period of time. Instead of being able to play every sport in the world, he eventually learns that the sports he can play at all are all very different, and he ends up having very limited skill in many different sports.

The key to building successful learning is to focus on developing one skill at a time, with the goal of learning to execute the skill effectively as soon as possible in the real world.

- **USING EXISTING CODE AS A "BLACK BOX"**

This is a common pitfall for many people who begin learning to code. They think they have all the answers and are content to leave their solutions as "black boxes," ready for them to figure out.

However, when you start learning to code, you need to make decisions about your code. You need to know what your code is going to do, what it's going to look like, how you're going to test it, and so on. You should be making these decisions over time, with the guidance of a structured approach that helps you prioritize your choices, while also allowing you to learn as you go.

That said, there are things you can do to help you learn more quickly.

Don't use existing code unless you understand it well.

Even if it seems like a good idea, you should avoid someone else's code unless you can be certain you know what you're doing. Of course if you want to prototype something quickly that's the fastest way to go, but projects in production stage should be well-reviewed and understood. Use your own names for things. When you write code in a new project, you should have a good understanding of what's going on inside of other projects.

- **LACKING CONSISTENCY IN STUDYING**

This one applies to everyone — data scientists, software engineers and students. It is easy to get distracted by news, social media, YouTube. We study for some time then we give it a break for a week or the next 2 months. The rest of the time, we're looking at a ton of new stuff. This is not a big deal if you're just starting out, but it is if you're trying to become an effective data scientist. If you don't have a plan for how to move forward in your studies and become an effective data scientist, you are not going to be successful.

There are two parts of this mistake:

The first is that it's too easy for beginners to give up. They're busy, so why bother? The second part is that they have their heads in the sand. They believe that data science isn't a real discipline because it's too hard. I'm going to argue that it isn't, and I'll give a specific and realistic set of tips on how to become an effective data scientist.

All in all you should remember a couple of things:
- having a plan is a must to organize your time effectively
- be persistent in studying one thing at time until you understand it
- don't give up too easily

Working on those things will allow you to grow as a data scientist and as a person.

- **NOT SPENDING ENOUGH TIME ON EXPLORING AND VISUALIZING THE DATA**

If you're a Data Scientist data is the key to understanding what is really going on.

Data is the thing that helps you understand the context and the way things work. **Data is the tool you use to get to the**

root cause of the problem. So the more data you have, the more you can understand your problems, to better solve them.

You need to give yourself time to play with the data — visualize it in different ways, label it, make it nicer.

There's an exciting part of visualization, which I call "the beauty of visualization." The beauty of visualization is that it's a form of problem solving. If you think of visualizations as "solving problems" rather than "finding information," you'll be in good company: "problem solving" is one of the defining characteristics of visualizations.

I believe that one of the primary advantages of visualization over other forms of "problem solving" is that it doesn't require us to make the final decision on which solution to choose. It allows us to see the whole picture, and thus be able to make a good decision based on the whole picture. And since there is no final decision, the final outcome is always "best."

- **GIVING TOOLS AND LIBRARIES PRECEDENCE OVER THE BUSINESS PROBLEM**

As a Data Scientist, you should be practical. It's always nice and intellectually challenging to think about theoretical aspects of data science, but in the end your job is to understand data in order to solve a business problem.

Data Science requires a great deal of experience and a lot of technical knowledge. The data you are working with is not always easy to work with.

If your goal is to build a new product, data is everything. It is literally the only tool that can be used to make this new product a success. You don't have time to develop an advanced framework or an advanced data modeling technique. You need to build your product. And this means you need to solve the business problem of your company.

As a data scientist you should be working with data every single day. The crucial question here is:

How are we going to determine which of a billion possible combinations of data inputs will create a given business output?

- **SHYING AWAY FROM DISCUSSIONS AND COMPETITIONS**

It's natural that we tend to avoid discussions sometimes, because we don't feel confident in our solutions or simply because we don't want to be challenged. But shying away from them is a big mistake as it leads to stagnation. You won't learn anyway this way. In order to learn you need to go out of your comfort zone.

This means taking risks and making a lot of mistakes, but that also means making mistakes that work. It means you have to be in the uncomfortable zone and take risks in order to learn. You can't just run away from problems and don't take the time to reflect on what you did wrong. That's what you are going to have to do in order to learn how to build working projects.

Don't wait to get some feedback on what you did — ask for it yourself. You might be surprised how receptive the feedback is!

You'll be surprised by how often people have questions about what you did or how to do it. Give yourself some time to practice. Practice is an amazing tool for any programmer.

You'll notice you become more comfortable and efficient as you start doing more and more. For me, I have a great way of practicing when I'm working on my own code. I write a few lines of my own code, test it out, write some more, and then repeat the process. Eventually, I just write down my code so I can follow along easily with it. Then I go back and start at the beginning.

Don't worry too much about being perfect. Don't sweat the details. Just get it done the first time. If you screw up on the first time, you can correct it and get back to where you were before.

In your first few times doing anything, you'll be a bit nervous. That's normal. Keep calm and carry on.

- **NOT WORKING ON COMMUNICATION SKILLS**

Solving a problem is one thing, but communicating it to the rest of the team is another. You need to work on your communication skills to convince others and explain to them what you've done and how does that fit the whole process.

There are a number of ways to achieve this, each with its pros and cons. One of the most effective ways is to have a dedicated communication channel for the team. For example, a Slack or Discord channel can be useful to communicate what your plan is for the project. You may even be able to use a dedicated channel on your workstation so everyone can talk to each other in private, saving them from the hassle of asking others for their input. Many teams also do stand-ups — quick meetups each day or each week to explain their results to each other. Those are taken in small groups of 5-10 people and can do marvels.

These can be very helpful in keeping people on track in a competitive environment.

But even without the "stand-up", people are more than welcome to come in on any time to ask questions and ask for assistance.

Being helpful to others is one of the best way to learn more yourself.

Conclusion

Every company values data science; it is highly unlikely that they will hire someone who doesn't have sufficient skills. This can be a negative point in case you go into the company already believing that you are the best data scientist for that company.

It's worth bearing in mind, however, that some teams don't really care about and/or find great candidates. Be careful! Good company will help you grow. Bad company will make you lose your motivation.

Data Science Books you should read

If you're learning Data Science then apart from reading open-source code on Github, Arxiv papers or course materials, a good idea is to read a Data Science book with actual coding examples, here's the list of Data Science books you should read.

Introductory level

If you're just starting your adventure with Data Science, you should definitely try:

Data Science from Scratch is what the name suggest: an introduction to Data Science for total beginners. You don't even have to know Python to start.

If you're a total beginner but you'd like to go more in Machine Learning direction from, *Introduction to Machine Learning with Python* is a book for you. It also doesn't assume you know Python.

Data Science Books you should read

Intermediate Level

If you've already read 1 or 2 Data Science books, done 1 or 2 projects for yourself and got accustomed to working with data a little bit, here are the books which will take you further.

Python for Data Analysis is the perfect way to get to know better standard Python libraries like NumPy or pandas. It is a complete treatise starting also from reminding you how Python works.

Python Data Science Handbook is a great guide through all standard Python libraries as well: NumPy, pandas, Matplotlib, Scikit-learn.

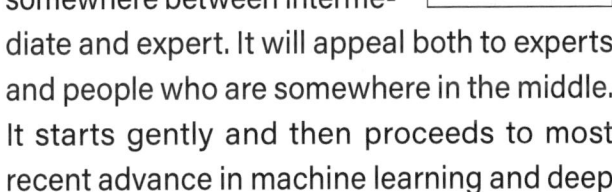

Python Machine Learning is somewhere between intermediate and expert. It will appeal both to experts and people who are somewhere in the middle. It starts gently and then proceeds to most recent advance in machine learning and deep learning. Great read!

Python for Finance is a must read if you're into finance and data science. It focuses on how to use data science tools to analyze financial markets and have many great examples illustrating that. It's very practical and will also appeal to people who don't work in finance on a daily basis.

57

Expert level

If you're approaching expert level, then actually reading scientific papers often makes more sense than reading books. However it is also time you study and implement deep learning in your solutions to go beyond the classical statistics. Three fantastic and now standard references are:

Deep Learning with Python was written by a creator of Keras, one of the most popular machine learning libraries in Python. The book starts gently, is very practical, gives pieces of code you can use right away and has in general many useful tips on using deep learning. An absolute must read in deep learning.

Deep Learning is an amazing reference for deep learning algorithms. It doesn't contain much code, but has great insights about how one should approach problems with machine learning, written by pioneers of deep learning.

If you're into mathematics, then you'll love *Machine Learning: a Probabilistic Perspective*. It's a tour-de-force through mathematics behind all machine learning methods. You probably won't be able to read it at once, but it's very useful as a reference in machine learning research.

Should you do a PhD in Data Science?

In one of the final chapters let's consider the ever living question: academia or business?

If you're considering doing a PhD in Data Science, you're going to have to decide whether you want to use your PhD to get a job as a data scientist, or to engage deeper in research and try to solve really complex, theoretical problems. This can be a challenging choice.

Is PhD for you?

Another thing is, whether a PhD is for you at all. In order to choose well:
- make sure data science is your passion and you're willing to spend an enormous amount of time on it
- make sure that you'll be able to fund your research
- be constantly curious and eager to learn about new things
- be persistent and don't easily give up

If you've decided it's for you, but you still wonder whether to do it at all, here's a couple of things to wonder:

- do you want to stay in academia afterwards and do research?
- do you want to go to a corporation or a start-up?
- do you want to become a freelancer, consultant or start your own business?

Depending on which of those three you choose, you'll enter a graduate school with a different mindset.

In the first case, you should focus on research as much as possible. In the second case, apart from research you should think about networking outside of academia and start spreading news about what you do in a form accessible to a general audience. In the third case you will need to polish your soft skills — communication, strategy, planning — in order to use your time effectively.

What is a PhD in Data Science?

Now to the PhD — doing a PhD in Data science can mean a couple things:
1. PhD in statistics at an economy faculty
2. PhD in mathematics at applied mathematics faculty
3. PhD in computer science or machine learning at computer science faculty

The answer which to choose will depend on how much coding hours you'll want to put in in the end and how much practical you want to get.

1. option is for those who want be practical but not code much

2. option is for those who don't want to be necessarily practical
3. option is for those who want to be practical and code a lot at the same time

There's no better or worse options here, it all depends on what you want to do after your PhD. Whether you imagine yourself coding for 8-10 hours a day? You want to meet people and visualize data? You want to use data science in business decisions? Answering those questions will let you decide.

The key here is to choose the ones that are most relevant to you.

There is a life after PhD

When you finish your research, you can look forward to a future of full time and part time jobs, or a life of consulting for a consultancy, doing data analytics for clients, or working in a company that's involved in data analysis.

The best decision for you may be to find a job that fits your strengths, and use your PhD to help you achieve your goals, whether that's getting the best data science job in the industry, or getting more research experience.

You're ready

That's all when it comes to this guide. Thank you for reading this far. You now know everything needed to find a data science job. What remains is honing your technical skills. Read the last chapter with Data Science Crash Course, open your Jupyter Notebook, and start coding along the way.

Build a couple of open-source projects on Github, with data from Kaggle.

Join Facebook groups for Data Science and start talking with more senior people about their data science jobs.

Soon, you'll find yourself starting your first data science job.

Good luck!

Data Science Crash Course

Let's learn Data Science in 2020

2020 is here and it's time to learn Data Science! This Data Science Crash Course is a quick way to start your journey with Data Science. Our goal is to start programming right away on whatever computer you have right now. Just by reading this text, you'll learn all the necessary terms to start coding. Let's go!

1. Introduction

- **WE USE PYTHON FOR DATA SCIENCE**

Python is a perfect language for beginners and experts alike, due to its popularity and clear structure. It's easy to pick up and thanks to community you'll be able to use it both for your first experiments as well as the most recent machine learning research.

- **OVERVIEW OF DATA SCIENCE CRASH COURSE**

First of all I'll talk about setting up your environment. In our case that means installing Anaconda on your computer, which will allow you to quickly run Jupyter Notebooks and there you'll be running short Python programs straight away from your browser. So from now until your first program you probably need like 15 minutes.

Then to do data science you will need a little bit of mathematics. I will review basic terms from linear algebra and statistics that you will need along the way. Those include:
- vectors and operations on them
- matrices and operations on them
- mean, variance, correlation
- random variables, dependence, conditional probability,
- Naive Bayes theorem

Then we can go directly to processing data:
- importing files in .txt, .json, .xml, .html
- storing data in NumPy arrays, pandas DataFrames and other structures (lists, dictionaries)

- **GETTING AND PROCESSING DATA**

I will also discuss how and where you can get interesting data from. Because Data Science is about Data, so you should find a dataset which is exciting to you. Some examples include:
- books and documents, for example from Project Gutenberg
- scraping websites (downloading sport or finance news)
- Kaggle databases
- Twitter API

- **STANDARD TECHNIQUES IN DATA SCIENCE**

Having data is the first step to Data Science and once we have data we can start working with. Standard techniques include:

Classification: we have 2 or more categories (labels) and we want to classify objects according to these labels. We talk here about supervised learning and standard techniques include: KNN, Naive Bayes, Decision Trees, XGBoost, neural networks.

Clustering: we don't have any labels but still we want to classify our data. We cluster our data by different measures of similarity, thus creating labels. We talk here about unsupervised learning and standard techniques include: k-means and DBSCAN.

Neural Networks is a separate topic in itself. The basic idea behind them is simplifying a problem into smaller pieces which can be processed separately by 'neurons'. I'll cover basic frameworks like Keras and Tensorflow, and then also discuss MNIST and other problems which are great to start with.

Dimensionality Reduction is a must if you want to visualize data or better understand it. Sometimes you just want to plot your 4D data in 2D and then you have to transform it to still capture the essence of information. Standard techniques include PCA and SVM.

Visualizing Data is important when it comes to business applications of Data Science. In the end you want to convey your findings to others, and the best way to do that is by presenting them with a nice graph, where everything is clear. We'll talk about plotly and Dash.

After this Crash Course you'll be prepared to tackle basic Data Science problems and learn more by yourself.

2. Anaconda and Jupyter Notebooks

Anaconda is a free, open-source distribution of both **Python** and R programming languages for data science and machine learning applications. It aims to simplify package management and deployment.

Ok, that might sound complicated but the truth is, it's all about giving you a framework where you can code. You need a compiler to run Python code, think text editor which 'runs' your program, and Anaconda gives you that plus much more. It gives you all packages you might want to use and Jupyter Notebooks.

But let's start with installation. Let's go to www.anaconda.com and download the most recent version:

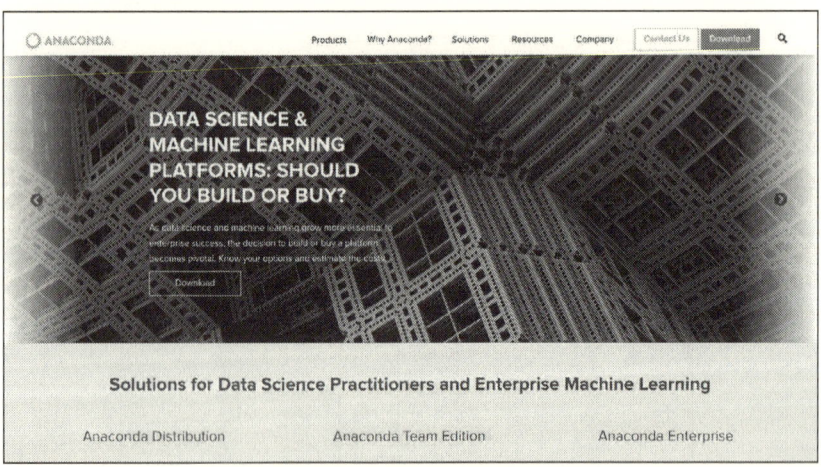

Anaconda website

Now that you have it, it's time to get into more explanations how it all works.

If you open Anaconda now you'll see this screen:

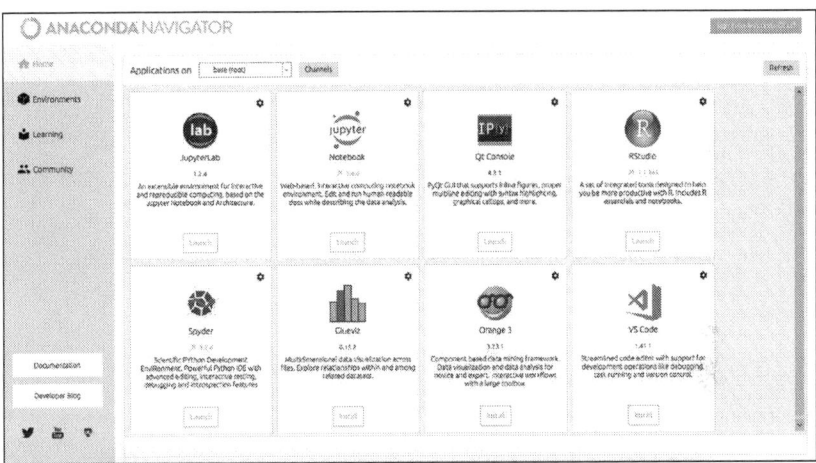

Anaconda Navigator

- **JUPYTER NOTEBOOKS**

Let's now go to Jupyter Notebooks. We can forget for now about the rest. Let's open Jupyter Notebook, you'll see the following screen:

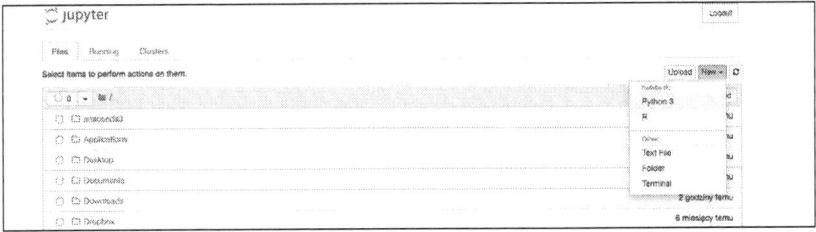

Jupyter Notebook

Click in the right corner on New and choose Python3. You'll see

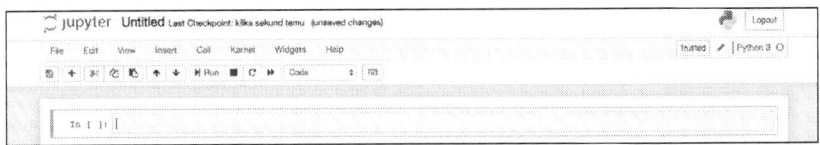

Jupyter Notebook

and... you're good to go! This is the command line where you can start writing code in Python and then execute by clicking Run.

Jupyter Notebooks are extremely handy way of writing code and keeping track of different experiments. It's not suited well for bigger projects, but for running quick experiments with data or training neural networks, it's perfect.

You usually start by importing packages/libraries by 'import ....'. Packages are ready-to-use pieces of code which allow you to save time by not having to write everything yourself. There are packages for everything:
- importing data (json)
- processing data (pandas, NumPy)
- text analysis (NLTK)
- neural networks (Keras, PyTorch, Tensorflow)
- visualization (Dash)

So it's often the case that before you do everything yourself you should check whether there's a relevant package already available. Of course sometimes you'll want to do it yourself anyway, just to learn.

On the ending note, JupyterLab is also a great tool, it's basically just a way to manage multiple Jupyter Notebooks at one time and thus it's very useful if you want to organize your code.

Now you have everything you need to write your first Data Science experiment in Python.

3. Linear Algebra and Statistics

Now we're going to review mathematics needed for Data Science.

- **LINEAR ALGEBRA FOR DATA SCIENCE**

Linear algebra is all about manipulations with vectors and matrices. It's both notation and useful way of manipulating object. Vector at its core is just a tuple of number written in a column (or a row):

$$y = \begin{bmatrix} x_1 \\ x_2 \\ \vdots \\ x_m \end{bmatrix}$$

You can perform operations on vectors like adding by adding each respective term—they need to have the same length. You can multiply a vector by a scalar, that is a real number, by multiplying each of the entries by this real number.

Now usually vectors are considered in certain larger space, for example like [1,2] on a plane, where they denote coordinates. Suddenly it becomes all very visual. Also it becomes evident that you can measure distance between vectors (it's a distance between points on a plane) or try to compute angle between two vectors.

It's really useful in Data Science when you write your data as vectors and then perform operations on them in order to measure them. Linear Algebraic methods are necessary to do that.

Vectors generalize to matrices, n by m arrays, which has n times m entries written in rows and columns:

$$A_{m,n} = \begin{pmatrix} a_{1,1} & a_{1,2} & \cdots & a_{1,n} \\ a_{2,1} & a_{2,2} & \cdots & a_{2,n} \\ \vdots & \vdots & \ddots & \vdots \\ a_{m,1} & a_{m,2} & \cdots & a_{m,n} \end{pmatrix}$$

Again you can add matrices of the same shape, multiply them by a scalar and perform other operations. For example you can multiply matrices of certain compatible shapes. This is crucial for many applications in Data Science, so you definitely should practise multiplying matrices before going any further.

That'll be all you need from Linear Algebra in the beginning. At some point it's worth reading about tensors which are a generalization of matrices to higher dimensions.

- **STATISTICS**

Statistics is about collection, organization, displaying, analysis, interpretation and presentation of data. That comes together with probability with theory, because you'll need to model unknown data by sampling it from certain distributions.

Crucial concepts are variance and bias. Following Wikipedia:
- The bias error is an error from erroneous assumptions in the learning algorithm. High bias can cause an algorithm

to miss the relevant relations between features and target outputs (underfitting).
- The variance is an error from sensitivity to small fluctuations in the training set. High variance can cause an algorithm to model the random noise in the training data, rather than the intended outputs (overfitting).

In other words bias is related to having too small of a sample of data compared to the whole set and variance is related to being too focused on each small change in our data. There's a trade off between those two concepts and it's often a case how much bias we trade for how much variance.

In order to start operating with machine learning models, you need to learn the basics of probability theory like

- random variables
- conditional probabilty
- naive Bayes theorem

Those concepts are crucial for understanding the basic mathematics behind Data Science and Machine Learning. The best way to learn them is do a couple of exercises and compute actual probabilities and matrices in particular cases. There are plenty of books and materials online so I won't copy that here.

4. Processing Data

Let's finally do something in Python with Data. I'll review basic techniques for processing data. How can you store information.

We've already learned we want to represent our data as vectors and matrices.

- **IMPORTING DATA**

We can start by importing files. You might have on your computer already

- txts
- Excel spreadsheets
- jsons
- xmls
- csv

And you can put it all to work, by importing it into Jupyter Notebook.

Most of it is done with 'open (...)' for example like this:

```
import csv

with open('example.csv') as csvfile:
    readCSV = csv.reader(csvfile, delimiter=',')
    for row in readCSV:
        print(row)
        print(row[0])
        print(row[0],row[1],row[2],)
```

Open CSV in Python

Have a look at Python Programming or Real Python to read more about it.

The same goes for other forms of files.

We already know that we want to represent our data in form of arrays (matrices, vectors), so let's see how we can do it in Python.

- **STORING DATA**

Now the question is how to store them. There are a couple of standard ways to do it.

Numpy Arrays is an easy way to represent arrays and NumPy is one of the best libraries for doing Data Science. Have a look at an official documentation of NumPy. And here's a snippet from Jupyter Notebook if you want to define a vector [1,2,3]:

```
[1]: import numpy as np
     np.array([1, 2, 3])

[1]: array([1, 2, 3])
```

NumPy Arrays are a great Data Science Tool

Quick and simple, right?

Pandas is an open source, BSD-licensed library providing high-performance, easy-to-use data structures and data analysis tools.

DataFrame, a spreadsheet tabular format, is a part of Pandas and also allows to play with Data. This example shows how to use it to create a simple spreadsheet:

```
[2]: import pandas as pd
     d = {'col1': [1, 2], 'col2': [3, 4]}
     df = pd.DataFrame(data=d)
     df
```

[2]:
	col1	col2
0	1	3
1	2	4

Pandas DataFrame for Data Science

It's really efficient and it's great to use, especially if you have used extensively Excel before, this will all be very familiar.

Apart from those two very efficient packages, Python itself is coming with plethora of data structures. Just to number a couple which comes in handy when it comes to manipulating data:
- lists
- dictionaries
- tuples

The best way to learn is to play around with it, so open your Jupyter Notebook now!

5. Getting Data

Data is crucial for doing Data Science. Often we have to clean it first in order to start using it. Let's discuss how to do it.

If you don't have any interesting data on your computer, then the best way is to just scrape information from the web. It's pretty easy with Python with packages like 'requests' and 'BeautifulSoup' (to clean data).

Most websites are easily scraped using requests and it's all just a matter cleaning.

- **SCRAPING WEB FOR DATA SCIENCE**

You can download useful data from the web. We're living in digital world, where virtually anything is available online. Most of data can be accessed from different websites by scraping. This way of getting data is very satisfactory but takes time, because you have to clean HTML code along the way.

Especially financial data is great to get. You can analyse it with tools like Pandas DataFrames and NumPy Arrays which we discussed in the previous section.

Let's say you want to get information from Nasdaq. With GET you can get HTML code right away:

```
import requests
html = requests.get('https://nasdaq.com')
html.text
```

'<!DOCTYPE html>\n<html lang="en" dir="ltr" prefix="conten
c/terms/ foaf: http://xmlns.com/foaf/0.1/ og: http://ogp
p://schema.org/ sioc: http://rdfs.org/sioc/ns# sioct: ht
re# xsd: http://www.w3.org/2001/XMLSchema# ">\n <head>\n
title" content="Daily Stock Market Overview, Data Updates,
content="homepage" />\n<meta itemprop="acquia_lift:page_ty
nguage" content="en" />\n<meta itemprop="acquia_lift:conte
_keywords" content="" />\n<meta itemprop="acquia_lift:post
tent="5ded78c5-4bf5-4a7b-ba13-a0cce7c580a3" />\n<meta item

Requests Python for Data Science

However as you can see this is far from readable. The next step would be to use BeautifulSoup to make it pretty:

```
from bs4 import BeautifulSoup
soup = BeautifulSoup(html.text, 'html.parser')

print(soup.prettify())
```
```
<!DOCTYPE html>
<html dir="ltr" lang="en" prefix="content: http://purl.org/rss/1.0/module
tp://xmlns.com/foaf/0.1/  og: http://ogp.me/ns#  rdfs: http://www.w3.org/
oc: http://rdfs.org/sioc/ns#  sioct: http://rdfs.org/sioc/types#  skos: I
w.w3.org/2001/XMLSchema# ">
 <head>
  <meta charset="utf-8"/>
  <meta content="Daily Stock Market Overview, Data Updates, Reports &amp;
  <meta content="homepage" itemprop="acquia_lift:content_type"/>
  <meta content="node page" itemprop="acquia_lift:page_type"/>
  <meta content="en" itemprop="acquia_lift:context_language"/>
  <meta content="home" itemprop="acquia_lift:content_section"/>
  <meta content="" itemprop="acquia_lift:content_keywords"/>
```

BeautifulSoup for Data Science

This is more readable but still you need to put in the work to extract some informations from it. That's a great way to start though.

Have a look at BeautifulSoup documentation for more details.

Scraping is great, because you can learn Data Science by playing around with more information about your hobby. Say if you like video games and want to analyse different statistics then scraping is the best option. You'll learn more about particular websites of interest to you.

- **DATASETS AVAILABLE ON THE INTERNET**

Other way to get data is to access already prepared datasets. There countless of examples of them and I'll just discuss some major datasets here.

Project Gutenberg is a great source for downloading books. You can access older books for which copyrights expired. Think Shakespeare.

Twitter API is another fantastic source. Register a Developer account and then you can start playing around with automation. For example you can pull data about a single hashtag and how people are responding to recent events. This is especially great for research in social sciences or marketing.

Kaggle (www.kaggle.com) is a fantastic resource for Data Science competitions but also for ready-to-use datasets which you can download directly, for free, from their website. Major companies often announce their competitions there so you'll see already what's of interest in a wider Data Science community.

- **ESTABLISHED DATASETS FOR DATA SCIENCE**

On top of that there are a bunch of datasets which are used for benchmarking machine learning models. I won't go into details here, but I want to just to direct you to a couple which simply are worth knowing:
- MNIST—a dataset of handwritten digits
- CIFAR-10–60,000 images with 10 classes
- ImageNet—an image database organized according to the WordNet hierarchy
- IMDB Reviews—a dataset of movie review from IMDB
- Wikipedia—great resource of texts, you can find a dataset from it online

That's all in this lecture. Now it's time to open your Jupyter Notebooks and play with your favourite dataset.

Practice is always the best way to learn Data Science.

6. Classification and Supervised Learning

We have already learned about storing data and where to get data from. Let's now cover standard techniques for classifying data, which is the basic application of Data Science.

- **WHAT IS SUPERVISED LEARNING**

Imagine you have data with labels attached. Think images of animals with a description whether it's a cat or a dog (classification problem). Another example is data about customers in an ecommerce with information like age group, occupation, past shopping (regression problem). Supervised learning deals with this type of problems, where you have labels attached to data so that you can 'supervise' the learning process of your algorithms using those labels as a guidance.

Classification is the problem of identifying to which label (category) a new object belongs, on the basis of a training set of data containing objects whose labels are known. Examples are assigning a given email to the "spam" or "non-spam" class, and assigning a diagnosis to a given patient based on observed characteristics of the patient.

Regression is a statistical process of estimating a value ('continuous label') based on other features or variables.

- **CLASSIFICATION AND REGRESSION ALGORITHMS**

K-Nearest Neighbours (KNN) is the most standard example for both classification and regression. We're looking at objects closest to a given example we start with and attach a label based on that. K stands for a number of neighbours we're looking at. Have a look at this implementation of KNN, when K=2, using sklearn, where we want to understand 6 points on a plane:

```
from sklearn.neighbors import NearestNeighbors
import numpy as np
X = np.array([[-1, -1], [-2, -1], [-3, -2], [1, 1], [2, 1], [3, 2]])
nbrs = NearestNeighbors(n_neighbors=2, algorithm='ball_tree').fit(X)
nbrs.kneighbors_graph(X).toarray()

array([[1., 1., 0., 0., 0., 0.],
       [1., 1., 0., 0., 0., 0.],
       [0., 1., 1., 0., 0., 0.],
       [0., 0., 0., 1., 1., 0.],
       [0., 0., 0., 1., 1., 0.],
       [0., 0., 0., 0., 1., 1.]])
```

KNN with sklearn

Naive Bayes is a standard approach. You simplify a situation and assume that actions/objects in your data are independent (in a probabilistic sense) and hence you can compute probabilities using Bayes theorem. Formally "Naive Bayes methods are a set of supervised learning algorithms based on applying Bayes' theorem with the "naive" assumption of conditional independence between every pair of features given the value of the class variable." You can implement it using sklearn again, for example with this example using a standard dataset of irises and modeling conditional probability using normal distribution (GaussianNB):

```
from sklearn.datasets import load_iris
from sklearn.model_selection import train_test_split
from sklearn.naive_bayes import GaussianNB
X, y = load_iris(return_X_y=True)
X_train, X_test, y_train, y_test = train_test_split(X, y, test_size=0.5, random_state=0)
gnb = GaussianNB()
y_pred = gnb.fit(X_train, y_train).predict(X_test)
print("Number of mislabeled points out of a total %d points : %d"
      % (X_test.shape[0], (y_test != y_pred).sum()))
Number of mislabeled points out of a total 75 points : 4
```

Naive Bayes with sklearn

Regression is used when you try to predict a precise value of a label. So it's a continuous variant of a classification problem. This would be perfect for trying to determine an income of a person based on people living close to him.

Linear regressions is the simplest—where you assume your data can be modeled by a linear function. There are plethora of regressions modeled after various functions, the most popular being sigmoid and ReLu (rectified linear function).

Again Linear Regression is easy to implement with sklearn.

Another classification technique is decision trees, where you try to have a Q&A format of data like:
- does this person live in a flat or a house?
- is this person over thirty?
- does this person have kids?

Trees can be built automatically. For example you can build a simple decision tree to divide the plane into two regions using sklearn:

```
from sklearn import tree
X = [[0, 0], [1, 1]]
Y = [0, 1]
clf = tree.DecisionTreeClassifier()
clf = clf.fit(X, Y)
clf.predict([[2., 2.]])

array([1])
```

Decision Trees with sklearn

Building trees is great because then you can join a couple of trees into a larger model via voting. Decision of each decision tree is taken into account and all decisions are weighted (arithmetic mean in the simplest case). This kind of methods are called ensemble learning.

- **MORE ADVANCED ALGORITHMS**

XGBoost is another standard technique, worth reading about. Official Documentation reads: "XGBoost is an optimized distributed gradient boosting library designed to be highly efficient, flexible and portable. It implements machine learning algorithms under the Gradient Boosting framework. XGBoost provides a parallel tree boosting (also known as GBDT, GBM) that solve many data science problems in a fast and accurate way. The same code runs on major distributed environment (Hadoop, SGE, MPI) and can solve problems beyond billions of examples." You can think about XGBoost as more advanced decision trees.

Neural networks are the ultimate tool often, when it comes to supervised learning. You build an architecture which is able to learn labels based on past examples. I will talk about them soon but first I'll explain what to do when you don't have labels a priori, that is how one go about unsupervised learning or clustering.

7. Clustering and Unsupervised Learning

We learned about supervised learning and what to do when you have a dataset with labels. Let's now look at datasets with no labels provided and talk about unsupervised learning.

- **WHAT IS UNSUPERVISED LEARNING**

Imagine we have raw data like social statistics related to marketing. For example you're trying to understand who has bought a

MacBook from your ecommerce and you'd like to find people who are similar. Or you're selling tickets through an online platform and you try to group your clients into different categories so that you can have a coherent message to each group.

In order to cluster data or group your data into categories (which are not given a priori!), you have to use one of clustering algorithms. Again sklearn will come helpful. Let's review two basic methods with a code example from sklearn.

- **CLUSTERING METHODS IN DATA SCIENCE**

k-means is the basic technique in clustering. "K" here stands for a number of clusters you want to have. This is arbitrary, you choose this parameter, but there are methods (see elbow method for example), where you can automatically infer the best number of clusters. You can use sklearn to group the famous Iris dataset into groups. For example this would be the result of 3 cluster:

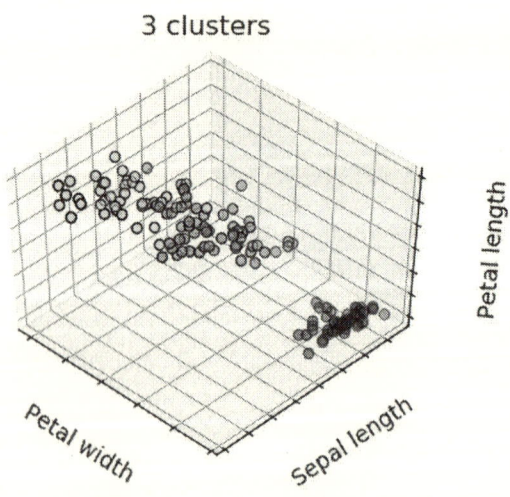

Clustering with k-means

If you want to have a simple k-means use case then you can simply cluster into 2 groups points on a plane.

```
from sklearn.cluster import KMeans
import numpy as np
X = np.array([[1, 2], [1, 4], [1, 0],
              [10, 2], [10, 4], [10, 0]])
kmeans = KMeans(n_clusters=2, random_state=0).fit(X)
kmeans.predict([[0, 0], [12, 3]])
```

array([1, 0], dtype=int32)

sklearn k-means

The key to k-means and other technique is having a metric, so a well-defined distance by which we can measure how similar are objects in our dataset. In most practical data coming from spreadsheets, metric is easy to define, we just take the usual distance between vectors in a space as data is purely numeric. Of course you can make it more involved, especially if your data is noisy or you are trying to extract really implicit data, but you don't have to think for example how to embed a word into a vector space.

Another Data Clustering Algorithms is Density-based spatial clustering of applications with noise or DBSCAN for short. Following sklearn documentation: "The DBSCAN algorithm views clusters as areas of high density separated by areas of low density. Due to this rather generic view, clusters found by DBSCAN can be any shape, as opposed to k-means which assumes that clusters are convex shaped. The central component to the DBSCAN is the concept of core samples, which are samples that are in areas of high density. A cluster is therefore a set of core samples, each close to each other (measured by some distance measure) and a set of non-core samples that are close to a core sample (but are not themselves core samples)." This is also one of standard techniques, it might sound complicated at first but the principle is easy:

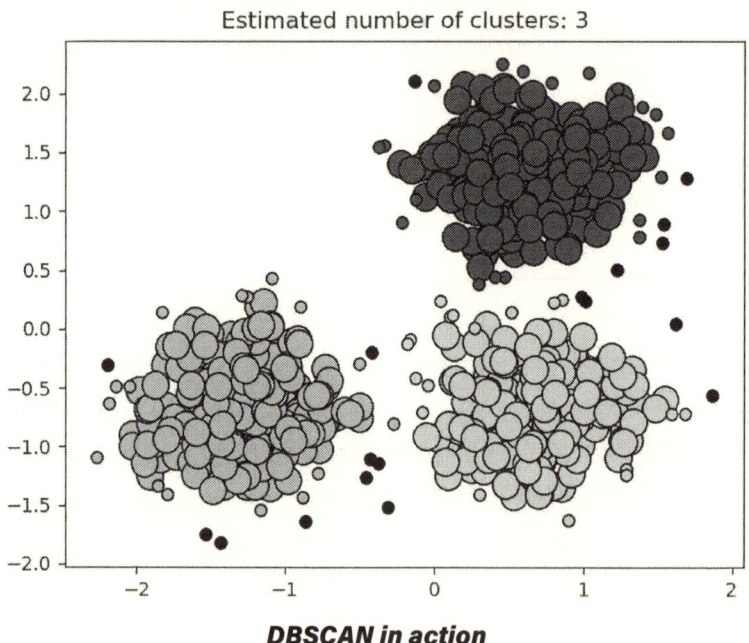

DBSCAN in action

If you want to know how to implement DBSCAN on an example, look for a Credit Card dataset done with DBSCAN.

- **DATA SCIENCE IS PRACTICAL**

The best way to learn is practice, so I highly recommend you open your Jupyter Notebook now, go on Kaggle and search for a database which you can use in experiments. There are great datasets in those examples above:
- Credit Card data
- Iris dataset
- Points on a plane

and you can find plenty of other datasets online, ready to use—both cleaned and raw.

Clustering often appears in nature and you're certainly will use it if you were to work for a larger organization like a bank or an insurance company. It's just natural that you are trying to group unlabeled data into categories which can be explained. Then if you managed to build categories by clustering, you're going back to classification problem with new data as you try to put new objects into already existing groups which came from clustering.

In the end Data Science is about classifying and clustering and extracting information from that. It just takes time to master.

8. Neural Networks

We're going to talk about Neural Networks and how to use them to classify data. It's going to be a gentle introduction to neural networks as I'm assuming you have never used them.

- **WHAT ARE NEURAL NETWORKS**

Neural Networks start with perceptrons, modeled on a single neuron in a human brain. You can think about it as a couple of functions composed one by one. You have the input data, then you apply an activation function—that can be a linear function, ReLu or a sigmoid function, or anything else—then you pass the data to the next node and again apply the activation function at a given node. It all goes in layers like that (this is more general feedforward network depicted):

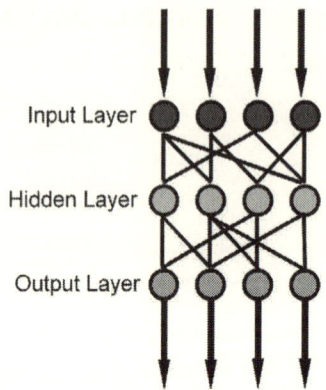

Feedforward Neural Network

So let's start with a perceptron example. We have a look at how they are implemented with sklearn, but then we switch to Keras framework. Keras is a high-level neural networks API, written in Python and capable of running on top of TensorFlow. TensorFlow is an end-to-end open source platform for machine learning. It has a comprehensive, flexible ecosystem of tools, libraries, and community resources that lets researchers push the state-of-the-art in ML and developers easily build and deploy ML-powered applications.

Here's a simple example with sklearn where we try to fit numbers using perceptron:

```
from sklearn.datasets import load_digits
from sklearn.linear_model import Perceptron
X, y = load_digits(return_X_y=True)
clf = Perceptron(tol=1e-3, random_state=0)
clf.fit(X, y)
clf.score(X, y)
```

0.9460211463550362

Sklearn Perceptron Neural Network

As you can see, code is short and simple, but details are in mathematics. At this point, you should go back to some linear algebra in order to understand what's going on.

The following paragraph is more technical, but I tried to put all the terms together (with links so you can read more about each concept):

Formally a multilayer perceptron is a class of feedforward neural network. A feedforward neural network is a neural network wherein connections between the nodes do not form a cycle. Backpropagation is an algorithm in training feedforward neural networks for supervised learning. Backpropagation computes the gradient of the loss function with respect to the weights of the network for a single input/output example. In this context it's also important to understand gradient descent, which is an algorithm used to find a local minimum of a function.

Let me note here that deep learning is machine learning with neural networks of at least 3 layers.

- **NEURAL NETWORKS IN KERAS**

Now let's go back to examples. As I've mentioned Keras and Tensorflow are most prevalent machine learning frameworks currently and Keras is especially easy to pick up. It's a bit like Lego when you build sequential models. Let's have a look at this example to see again Multilayer Perceptron this time built with Keras:

```python
import numpy as np
from keras.models import Sequential
from keras.layers import Dense, Dropout

# Generate dummy data
x_train = np.random.random((1000, 20))
y_train = np.random.randint(2, size=(1000, 1))
x_test = np.random.random((100, 20))
y_test = np.random.randint(2, size=(100, 1))

model = Sequential()
model.add(Dense(64, input_dim=20, activation='relu'))
model.add(Dropout(0.5))
model.add(Dense(64, activation='relu'))
model.add(Dropout(0.5))
model.add(Dense(1, activation='sigmoid'))

model.compile(loss='binary_crossentropy',
              optimizer='rmsprop',
              metrics=['accuracy'])

model.fit(x_train, y_train,
          epochs=20,
          batch_size=128)
score = model.evaluate(x_test, y_test, batch_size=128)
```

Keras Neural Network

I won't get into details here, but what is happening here:
1. We import Keras.
2. We generate some random numbers to have some data to train on and then test on (X,y train/test).
3. We build a sequential model but adding 3 layers. Two of the them have ReLu as an activation function and the last one have a sigmoid. We also use dropout but I won't discuss that here, though it's worth to read more about it.
4. We then compile and fit the model to our train data.
5. We evaluate the model on our test data.

All in all this is how neural networks and machine learning work in a nutshell:
1. Set up the stage by importing packages, frameworks and data (both to train and test).
2. Build a model.
3. Fit the model to train data.
4. Evaluate the model on test data.
5. Tweak model's parameters, tweak data, repeat.

As Data Science is practical and it's all about testing and trying, now it's your turn. Open your Jupyter Notebook and play with data. I also recommend you read about how to use Keras with MNIST dataset.

9. Dimensionality Reduction

Let's talk about how to reduce a number of dimensions in our data, so that it can be visualised and better understood.

- **PRINCIPAL COMPONENT ANALYSIS**

Imagine you want to plot data but your data has too many dimensions to do that right away. Here comes Dimensionality Reduction. These methods allow you to look only at certain dimensions which have the most relevant information for you.

Standard technique is PCA, Principal Component Analysis, which is looking at eigenvectors (singular values) and projecting to those which capture the most of data. So you can transform for example 4D into 2D, by projecting onto 2D space spanned by two largest singular values.

Let's look at a sklearn code for a simple PCA use with a NumPy array:

```
import numpy as np
from sklearn.decomposition import PCA
X = np.array([[-1, -1], [-2, -1], [-3, -2], [1, 1], [2, 1], [3, 2]])
pca = PCA(n_components=2)
pca.fit(X)
print(pca.singular_values_)

[6.30061232 0.54980396]
```

PCA with sklearn

Here we just use PCA with 2 components and we get singular values for our space. To really understand what's going on, you should read a more about how to use PCA with real world problems. There are plenty of tutorials and visualisations online.

- **DIMENSIONALITY REDUCTION IN DATA SCIENCE**

Dimensionality reduction is important because you want to get rid of unnecessary data, irrelevant columns in spreadsheets, in order to apply data science algorithms in the most efficient way. Less parameters are easier to control and manipulate.

There are many other methods used for dimensionality reduction. More advanced include for example Manifold Learning. Manifold learning is an approach to non-linear dimensionality reduction. Algorithms for this task are based on the idea that the dimensionality of many data sets is only artificially high. t-SNE is one of the most well-known technique of manifold learning and you can read more about it in sklearn documentation.

All in all, how you're going to reduce dimensions depend entirely on data you're working with. Sometimes it'll be obvious from the start that you should ignore a couple of columns or rows to get better results. Sometimes it'll be hard to see what is really important and then you'll have to try a bunch of dimensionality reduction algorithms to unwind the data and dependencies between objects.

But that is why Data Science is practical—there's no understanding without playing around with data.

10. Visualisation

Finally I'm going to discuss how to present a Data Science project so that it's appealing and instructive to others. In other words, let's talk about visualisation.

- **HOW TO VISUALISE IN PYTHON**

I start by writing that a well-documented Jupyter Notebook is perfect for a visualisation, especially if you want to show your work to someone with technical background. This is often what employers expect from you to send, when you do a test problem for a job interview.

Otherwise, if you're talking with non-technical people, you will need more visual ways to show what you have achieved like:
- plot data using plotly or matplotlib
- create a heatmap
- show statistics in a table

Here's an example of a simple plot using matplotlib:

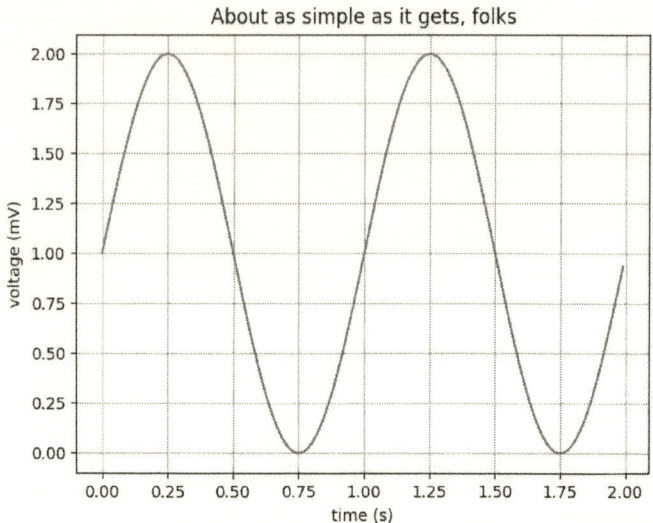

Matplotlib plot

The code for it is:

```python
import matplotlib
import matplotlib.pyplot as plt
import numpy as np

# Data for plotting
t = np.arange(0.0, 2.0, 0.01)
s = 1 + np.sin(2 * np.pi * t)

fig, ax = plt.subplots()
ax.plot(t, s)

ax.set(xlabel='time (s)', ylabel='voltage (mV)',
       title='About as simple as it gets, folks')
ax.grid()

fig.savefig("test.png")
plt.show()
```

matplotlib in Python

In general it's amazing what you can create with plotly. Below you can find a sample of visualisations from plotly, all done in Python:

Data Science Crash Course

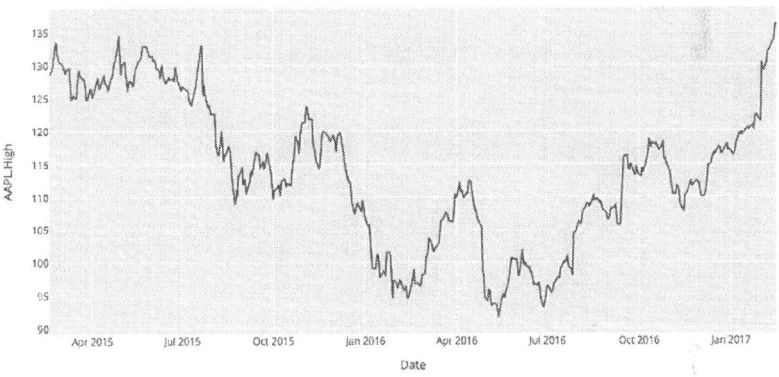

Simple Ternary Plot with Markers

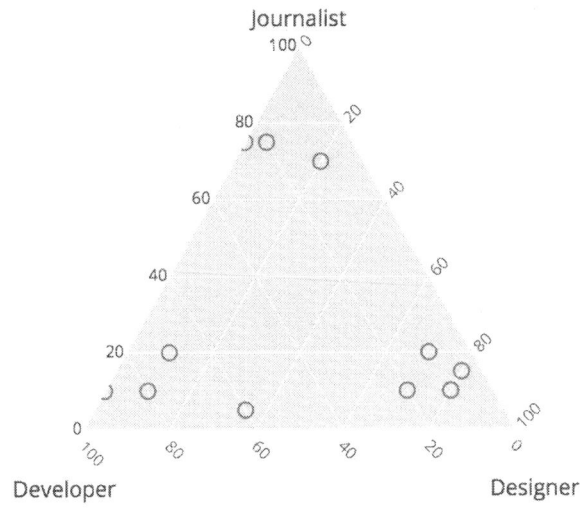

A good command of plotly will allow you to do really great things.

Going even further with plotly is Dash, by plotly, which allows you to build web-apps from Python. So it's really great to show to anyone really, whether a technical or non-technical. Have a look at official documentation of Dash to see what's possible. For example this interactive dashboard was done in Dash:

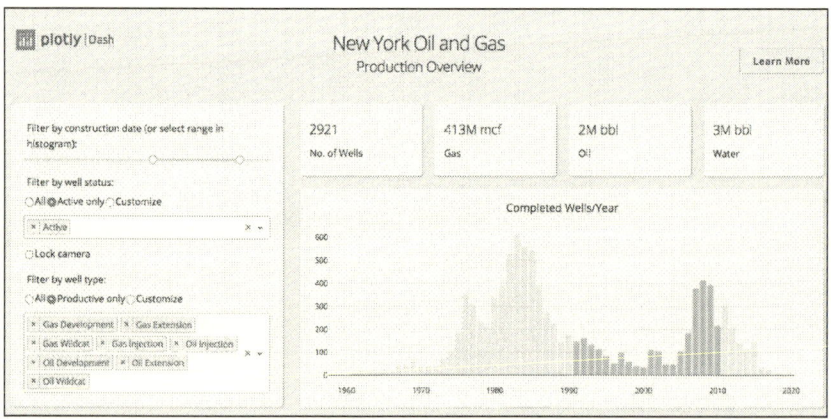

Dash dashboard in Python

As you can see, Python gives you amazing options for visualising and presenting your work. Now it's your turn to play with it in Jupyter Notebook.

- **SHARE YOUR OPEN-SOURCE PROJECTS**

The best way to share your projects is definitely Github. Upload your code with a well-written documentation and share it with your colleagues to get feedback. Of course you can't really do it, if you're working for a company, but if you're working towards your own open-source project then Github is a perfect way to share your work. Data Science has a great community so you're definitely going to hear feedback and learn more from others.

And that's all for this Data Science Crash Course.

I hope you've enjoyed it and learned useful things and you're ready to tackle some data science problems now.

Good Luck with your Data Science journey!

Made in the USA
Columbia, SC
04 August 2020